HARCOURT SCIENCE

VIRGINIA SOL SUPPORT FOR STUDENTS

GRADE 1

Harcourt

Orlando   Austin   Chicago   New York   Toronto   London   San Diego

Visit *The Learning Site!*
www.harcourtschool.com

Reviewers

Sharon Bowers
Kemps Landing Magnet School
Virginia Beach, Virginia

Brenda Dorman
Coleman Place Elementary
Norfolk, Virginia

Coleen Matthews
Coleman Place Elementary
Norfolk, Virginia

Diane C. Tomlinson, Ed. S.
Virginia Science Content Specialist/Codirector
Coalfield Rural Systemic Initiative
Lebanon, Virginia

Copyright © by Harcourt, Inc.

All rights reserved. No part of this publication may be reproduced or transmitted in any form or by any means, electronic or mechanical, including photocopy, recording, or any information storage and retrieval system, without permission in writing from the publisher.

Requests for permission to make copies of any part of the work should be addressed to School Permissions and Copyrights, Harcourt, Inc., 6277 Sea Harbor Drive, Orlando, Florida 32887-6777. Fax: 407-345-2418.

HARCOURT and the Harcourt Logo are trademarks of Harcourt, Inc., registered in the United States of America and/or other jurisdictions.

Printed in the United States of America

ISBN 0-15-338711-4

5 6 7 8 9 10 170 10 09 08 07 06 05

Contents

Lesson 1   **What People Need** 5
Science Standards of Learning 1.5, 1.5a

Lesson 2   **What Plants Are Like** 7
Science Standards of Learning 1.4c

Lesson 3   **Plants and Water** 11
Science Standards of Learning 1.7a

Lesson 4   **Animal Body Coverings** 13
Science Standards of Learning 1.5b

Lesson 5   **What Animals Are Like** 17
Science Standards of Learning 1.5c

Lesson 6   **Limited Natural Resources** 19
Science Standards of Learning 1.8

Lesson 7   **Air and Water Pollution** 23
Science Standards of Learning 1.8b

Lesson 8   **Water and the Weather** 27
Science Standards of Learning 1.7c

Lesson 9   **How Animals Change with the Seasons** 29
Science Standards of Learning 1.7b

Name _____ Date _____

LESSON 1

LEARN ABOUT

Vocabulary
shelter

What People Need

Needs

People need some things to stay alive, just as other animals do. We need food, water, and air to stay alive and grow. We also need safe places to live.

We need food to help our bodies stay healthy and grow. People also need water. We drink water to stay alive. The air around us also keeps us alive.

A **shelter** is a place that keeps people safe. People live in shelters to keep warm and dry.

■ How does the apple help this boy?

■ What does this girl need to breathe?

■ What does the girl get from the shelter?

SOL 1.5, 1.5a   Use with Unit A, Chapter 1.   Lesson 1 ■ Virginia SOL Support for Students ■ 5

Name _____ Date _____

Activity

Directions Make a chart to show what you need to live. Draw a picture of each thing. Tell where you find what you need.

\	What I Need to Live	
Need	**Picture**	**Where can I find this?**
Food		
Water		
Air		
Shelter		

Name _____ Date _____

LESSON 2

LEARN ABOUT

Vocabulary

edible   flowering
evergreen

What Plants Are Like

Some Plants Can Be Eaten

Some plants are food for other living things. Animals and people eat parts of plants. Plants that are eaten are called **edible** plants.

Some plants are not edible. They should not be eaten. They may have parts that are harmful. If you eat them, you could get sick.

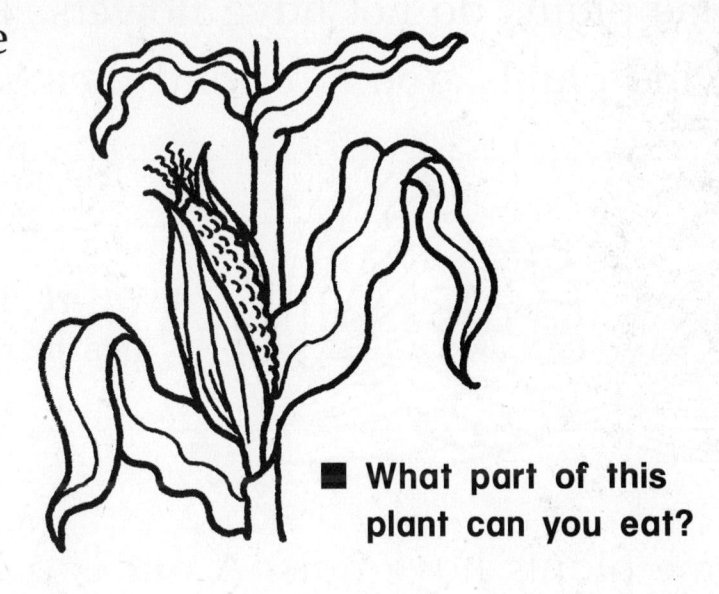

■ What part of this plant can you eat?

■ People should not eat these berries. What might happen if someone eats them?

SOL 1.4c Use with Unit A, Chapter 2.

Lesson 2 ■ Virginia SOL Support for Students ■ 7

Name _____ Date _____

Plants and Flowers

Many plants have flowers. These plants are called **flowering** plants. Their flowers make the seeds.

■ What part of this plant makes the seeds?

Some plants do not have flowers. They are not flowering plants. You may find their seeds in cones.

■ Where are the seeds on this plant?

Some plants have nuts. A nut is a kind of seed.

■ What is an acorn?

Name _____   Date _____

Some Plants Change

Some plants change when seasons change. They have green leaves in spring. Their leaves change color in fall. In winter, their leaves fall to the ground.

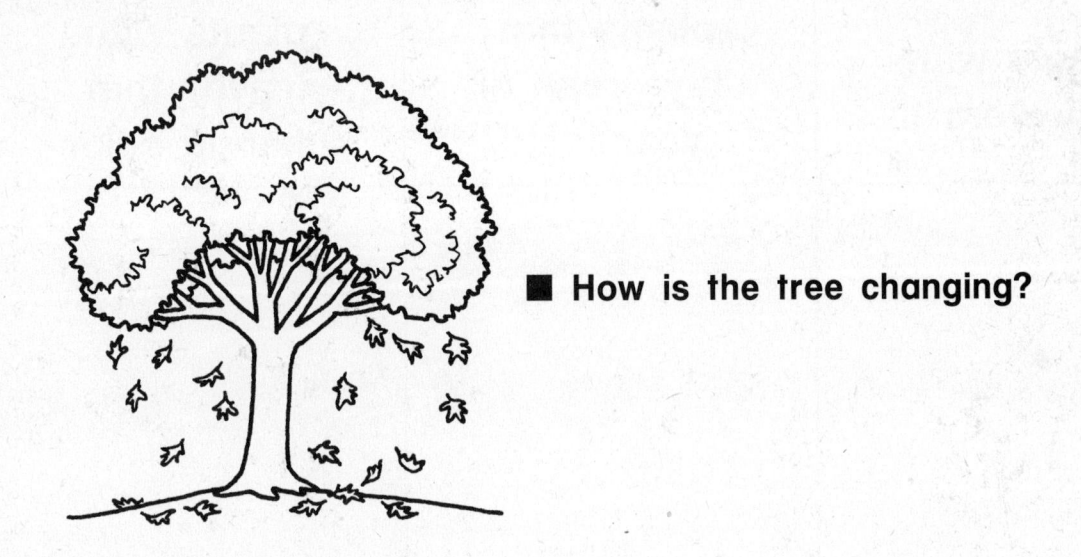

■ How is the tree changing?

Some plants with needles stay green all year long. A plant like this is called an **evergreen.** The needles do not change color unless the tree is sick.

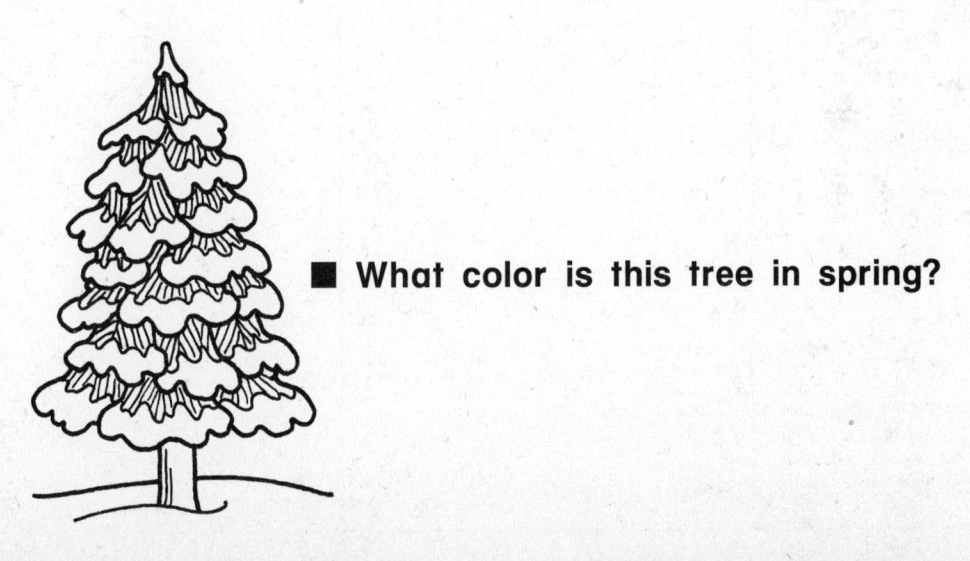

■ What color is this tree in spring?

Name _____ Date _____

Activity

Directions Find pictures of flowering plants, plants with needles, and plants with leaves. Classify the plants in a chart.

plants with flowers	plants that stay green all year long	plants with leaves that change color

Name _____ Date _____

LESSON 3

LEARN ABOUT

Vocabulary

wilt

Plants and Water

Plants Need Water

Plants need water to stay alive. Roots carry water to the stems. Water helps the stems hold up the plant.

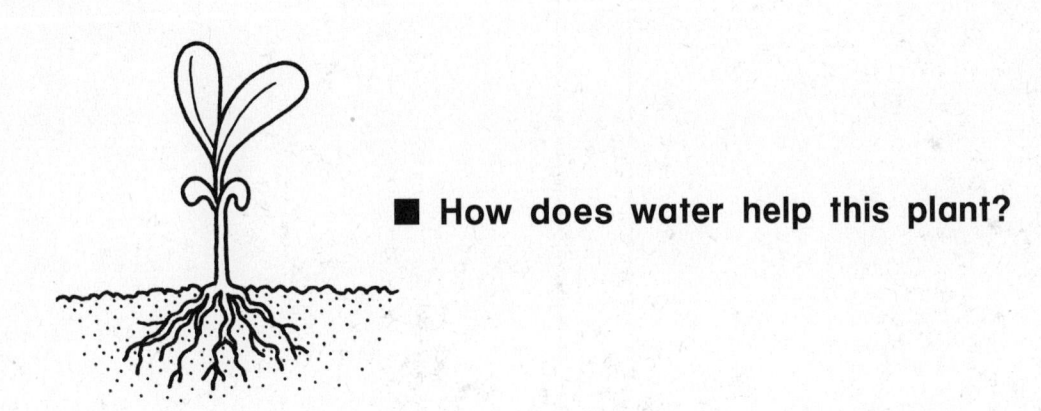

■ How does water help this plant?

A plant will not grow without water. It will **wilt,** or start to fall over. The stems will not hold up the plant. The plant becomes dry. The green leaves turn brown. The plant starts to die.

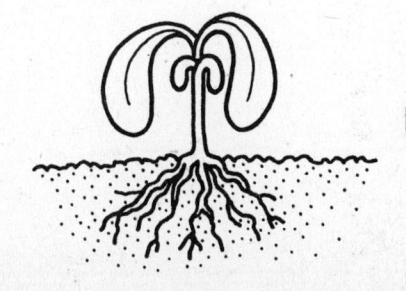

■ Why did this plant wilt?

SOL 1.7a Use with Unit A, Chapter 2.

Name _____ Date _____

Activity

Directions Draw what each garden will look like.

These garden plants were watered.

These garden plants were not watered.

12 ■ Lesson 3 ■ Virginia SOL Support for Students

SOL 1.7a Use with Unit A, Chapter 2.

Name _____ Date _____

LESSON 4

LEARN ABOUT

Vocabulary
shells

Animal Body Coverings

Body Coverings Are Different

Body coverings protect animals. Animals have different body coverings. Animals that have the same kind of body covering can be grouped together.

Fur or Hair

Mammals have bodies that are covered with fur or hair. The fur or hair keeps them warm in winter.

■ How does fur help this polar bear?

Some animals spend all day in the sun. The sun can burn skin. Fur or hair protects animals from the sun.

SOL 1.5b Use with Unit A, Chapter 3. Lesson 4 ■ Virginia SOL Support for Students ■ 13

Name _____ Date _____

Feathers

Birds are covered with feathers. Feathers help keep birds dry.

■ How do feathers help this duck?

Feathers help birds in other ways. Some feathers keep birds warm. Penguins have feathers to help keep them warm. Other feathers help some birds fly.

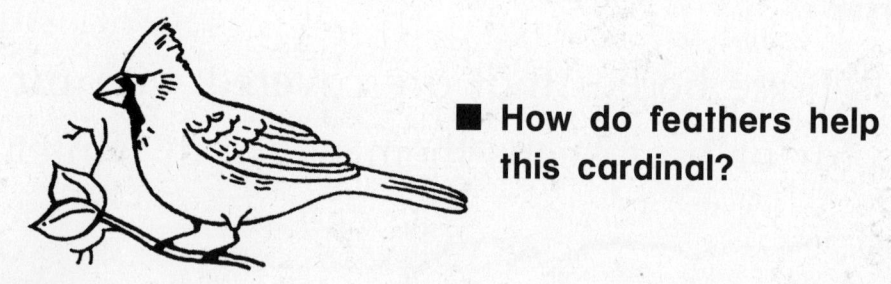

■ How do feathers help this cardinal?

Scales

Some animals have scales that cover their bodies. Reptiles are animals that have dry scales. They live on land.

Fish also have scales. They live in water. Their scales are wet.

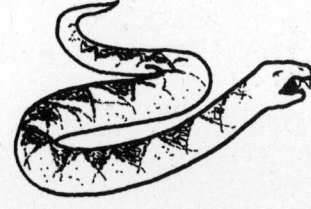

■ How are the animals alike? How are they different?

14 ■ Lesson 4 ■ Virginia SOL Support for Students   SOL 1.5b Use with Unit A, Chapter 3.

Name _____ Date _____

Smooth, Wet Skin

Amphibians have smooth, wet skin. They live near water. Amphibians must keep their skin wet.

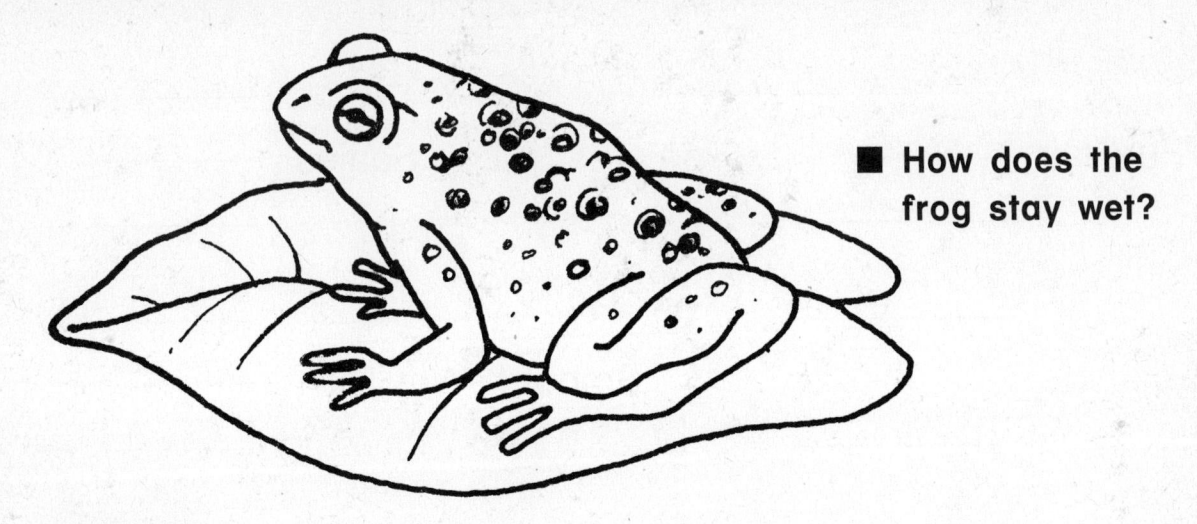

■ How does the frog stay wet?

Shells

Shells are another kind of body covering. Shells are hard. Animals with shells have soft bodies. Shells protect the animals. Snails, clams, and oysters are animals with shells.

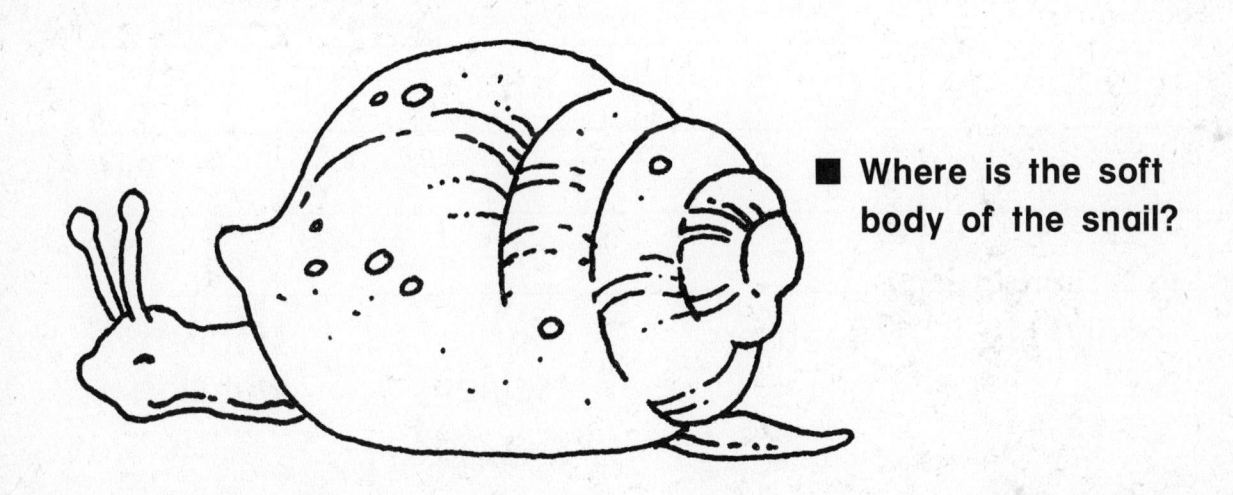

■ Where is the soft body of the snail?

SOL 1.5b   Use with Unit A, Chapter 3.   Lesson 4 ■ Virginia SOL Support for Students ■ 15

Name _____ Date _____

Activity

1 Count and write how many of each.

Fur/Hair _____   Dry Scales _____

Feathers _____   Smooth, Wet Skin _____

2 Color the graph to show your answers.

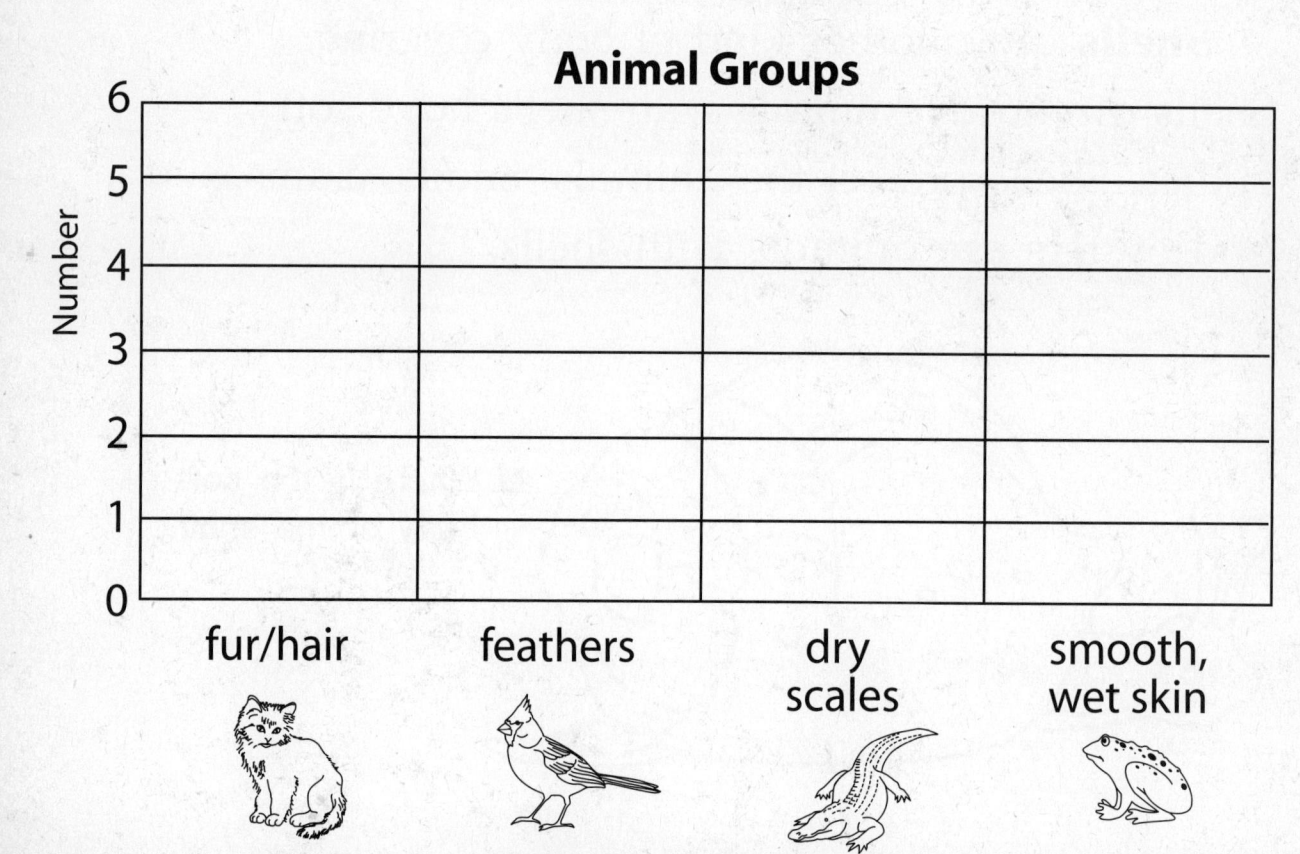

Animal Groups

Number (y-axis: 0–6)

Kind of Body Covering (x-axis): fur/hair, feathers, dry scales, smooth, wet skin

16 ■ Lesson 4 ■ Virginia SOL Support for Students   SOL 1.5b Use with Unit A, Chapter 3.

Name _____ Date _____

LESSON 5

LEARN ABOUT

Vocabulary
wild   tame

What Animals Are Like

Land and Water Homes

Animals live in different places. Some animals live on land. Some animals live in water.

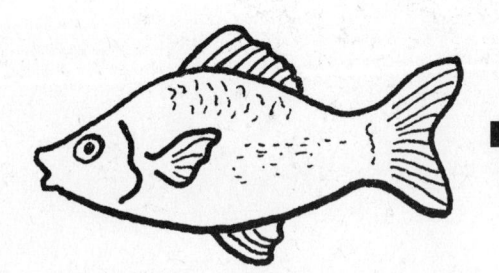

■ Where does this animal live?

Animals That Are Not Good Pets

Many animals are **wild.** Most wild animals live outside. Some are kept in zoos. Wild animals do not make good pets.

polar bear

Animals That Make Good Pets

Some animals are **tame.** They can live near people. Many dogs and cats are tame animals. Tame animals make good pets.

SOL 1.5c Use with Unit A, Chapter 3.   Lesson 5 ■ Virginia SOL Support for Students ■ 17

Name _____   Date _____

Activity

Directions Find pictures of animals. Classify your pictures to make a chart.

Land Animals	Water Animals

On your own paper, draw a picture of one tame animal and a picture of one wild animal. Label your pictures with the names of the animals.

Name _____ Date _____

LESSON 6

LEARN ABOUT

Vocabulary

limited

Limited Natural Resources

Using Natural Resources

People use land, water, and air. They also use plants and animals. People use these and other natural resources in many ways.

Water is a natural resource that people use for drinking, cooking, and washing.

■ How is this girl using a natural resource?

Forests are also a natural resource. People cut down trees to build houses. They use the wood to make tables and chairs.

SOL 1.8 Use with Unit C, Chapter 2.

Name _____ Date _____

Replacing Natural Resources

Some natural resources can be replaced after people use them.

Water is replaced when it rains. Rain fills up rivers and lakes. This can take a long time, so people must use water carefully.

■ How will water in this river be replaced?

Trees and other plants can also be replaced. People are able to plant more. Trees take a long time to grow. People must use them carefully.

■ How can the trees in this forest be replaced?

20 ■ Lesson 6 ■ Virginia SOL Support for Students   SOL 1.8 Use with Unit C, Chapter 2.

Name _____ Date _____

Limited Resources

Many natural resources are **limited.** Once they are used, there will be no more. They cannot be replaced.

Metals are a natural resource. Zinc is a kind of metal found in minerals in Virginia.

Minerals are limited natural resources. They cannot be replaced.

■ Why is zinc called a limited natural resource?

Soil is another limited resource.

There is a lot of soil on Earth, but it can be washed away by rain. People must take care of it because no one can make new soil.

■ Why do we need to take care of soil?

SOL 1.8 Use with Unit C, Chapter 2.   Lesson 6 ■ Virginia SOL Support for Students ■ 21

Name _____ Date _____

Activity

Directions Cut out magazine pictures of natural resources. Glue them in the chart. Write <u>yes</u> or <u>no</u> to tell if each resource can be replaced.

Natural Resource	Can it be replaced?

Name _____ Date _____

LESSON 7

LEARN ABOUT

Vocabulary

waste

Air and Water Pollution

Air

The air on Earth is a natural resource. People, plants, and animals need clean air to stay healthy.

■ How is this animal using air?

Air can get dirty. The smoke that comes from cars can make air dirty. The smoke that comes from factories can also make air dirty. Air is not healthful to breathe if it becomes very dirty.

■ What is making the air dirty?

Name _____ Date _____

Water

All living things need clean water. Many animals live in water. People and animals drink water. People use water for cooking and washing. Plants need water to make food.

■ Why does this dolphin need clean water?

Many things can make water dirty. Water is not healthful for people, plants, and animals when it becomes very dirty. Waste makes the water dirty. **Waste** is anything that is thrown away and not used. Waste comes from factories, homes, and farms. Waste also comes from animals and people.

Rain washes waste from yards, streets, and farms into rivers, ponds, and lakes.

■ How did this river get dirty?

Name _____   Date _____

Keeping the Air and Water Clean

People can help keep the air clean. Some people walk instead of using a car. They know that cars make the air dirty.

■ How are these people helping to keep the air clean?

People can help keep water clean. They can get rid of waste the right way. Throwing trash in the water makes it dirty.

■ How is this boy helping to keep water clean?

Name _____ Date _____

Activity

Directions

Circle things that make air and water dirty.

Look outside on a rainy day. Draw a picture to show where the rain goes.

Where does the rain go?

Name _____ Date _____

LESSON 8

LEARN ABOUT

Vocabulary

snow   sleet   hail

Water and the Weather

Water Falls to the Earth

Water drops fall to the Earth from clouds. When the weather is warm, they fall as rain. People wear raincoats or carry umbrellas when it rains.

Many places get very cold in winter. The cold air freezes the water drops in the clouds. The frozen drops of water fall as **snow.** People wear warm coats, gloves, hats, and boots when it snows.

Sometimes the water drops in the clouds fall as **sleet.** Sleet is made up of water drops that are partly frozen. Sleet looks like tiny pieces of ice.

Water drops sometimes fall as **hail.** In a thunderstorm, they may freeze in the clouds even in summer. They fall to Earth as small balls of ice. People should stay inside when hail falls so they won't get hurt.

sleet

SOL 1.7c Use with Unit D, Chapter 1.   Lesson 8 ■ Virginia SOL Support for Students ■ 27

Name _____ Date _____

Activity

Directions Draw pictures to show a person when there is rain, snow, sleet, and hail. Show what he or she should wear or where he or she should be. Tell what the air is like.

Rain	**Snow**
Sleet	**Hail**

Name _____ Date _____

LESSON 9

LEARN ABOUT

Vocabulary
shedding

How Animals Change with the Seasons

Some Animals Change

You wear clothes that help keep you cool in summer. You wear warm clothes in winter.

Animals also need to stay cool or warm. Some animals have body coverings that change with the seasons. These changes help animals stay alive.

Some animals get thicker fur when it gets cold. Fur keeps animals warm. This helps them stay alive in cold winters.

Some animals have dark fur that changes color in winter. Most of these animals live where it snows. Their fur changes to white. This helps them hide in the snow.

■ How does this bear stay warm in winter?

SOL 1.7b   Use with Unit D, Chapter 2.

Name _____ Date _____

Animals that get extra fur in winter lose it in summer. This is called **shedding.** Shedding helps keep animals cool.

■ How does this dog stay cool?

Activity

Directions What is an animal that changes its body covering in winter and in summer? Draw the animal to show how it changes.

My Animal in Summer	My Animal in Winter